wildfire

SCOTT THYBONY

WESTERN NATIONAL PARKS ASSOCIATION
Tucson, Arizona

This country has always burned.
And it needs to burn...

blowup

IN THE SUMMER OF 1910 THE WEST WAS ON FIRE. Crews of firefighters had struggled for weeks to contain hundreds of blazes sweeping across the Bitterroot Range, the hardest-hit sector. Then on August 20, at the moment they were getting the upper hand, gale-force winds struck. What happened next would haunt a generation of forest rangers. Waves of flame jumped rivers and mile-wide gorges on the border of Idaho and Montana, throwing firebrands far in advance of the inferno. Entire mountainsides, superheated by the approaching fires, ignited instantaneously. "The wind was so strong," said ranger Ed Pulaski who was rounding up firefighters from outlying camps, "that it almost lifted men out of their saddle." Trapped by swirling flames, they began to panic. "We were completely surrounded by raging, whipping fire....The whole world seemed to us men back in those mountains to be aflame. Many thought that it really was the end of the world."

ABOVE: Fort Apache Hotshots hike single file to the Meadow Creek Fire in Idaho.

ABOVE RIGHT: Firefighter uses a Terra Torch to light a prescribed burn.

LEFT: A Los Angeles City Fire Department helicopter drops water from it's 360-gallon tank in support of a linebuilding operation in Malibu, California.

Pulaski shouted above the terrific roar of the firestorm, ordering his forty-five firefighters to follow him. He knew the mountains and realized it was up to him to lead them to safety. As the day turned pitch black, each man placed his hand on the shoulder of the one in front and followed Pulaski. But the sudden blowup had cut off their escape. The only chance was to find, in all the smoke and fiery chaos, the War Eagle mine where a tunnel offered a degree of protection. Keeping close to Pulaski, they ran a gauntlet of flaming trees and searing blasts of heat. One man, dropping behind, was struck by a falling snag and killed. A bear that had lost its fear of humans joined the procession before disappearing in the smoke.

The fleeing crew reached the mine just before the full force of the firestorm hit. Dashing inside, Pulaski had them lie face down as the mine timbers at the entrance caught fire. One of them panicked and began to push his way out, forcing the ranger to draw his pistol and threaten to shoot the first man who tried to leave. "The men," he later wrote, "were in a panic of fear, some crying, some praying. Many of them soon became unconscious from the terrible heat, smoke, and fire gas."

Smoke filled the tunnel and Pulaski himself collapsed, lying unconscious for hours until the fire subsided and the air began clearing. By morning all but five of the men were still alive. The ranger, nearly blind, led the survivors across a charred landscape of smoking stumps and scattered flare-ups. Rags of scorched clothing hung from their bodies, and the intense heat had burned through their boots. All around them dead trees choked the slopes in vast, crisscrossing logjams. Crawling over the down trees and staggering along as best they could, Pulaski and his men made their way to the nearest town.

FAR LEFT: The Negrito Hotshots, a New Mexico crew, return to camp through a blackened forest.

LEFT: Still-burning ruins of a house at the forest edge

The big blowup of 1910 consumed three million acres of virgin timber, destroyed four towns, and took the lives of at least eighty-five people, mainly firefighters. Most of the destruction happened during a terrifying six-hour period. Stunned by the ferocity of the epic burn, forest officials began a policy of chasing every smoke and extinguishing every fire. Six hours of holocaust led to sixty years of aggrssive fire suppression.

For more than a generation, the voices of those who saw the role of fire as absolutely necessary to the survival of the forest ecosystem went unheard. And for more than a generation the conditions in the nation's forests and parklands grew increasingly volatile.

Over time, uniform stands of trees replaced the natural mosaic of mixed species and open parks. Dense thickets became susceptible to insects and disease, increasing the number of dying trees. A vast buildup of windfall on the forest floor covered a mat of dead leaves and needles ready to ignite when the weather turned hot and dry. In some forests, heavy grazing by livestock had already reduced grasses and the frequent light fires they fueled. Intervention in the natural cycle of fire had made the forests vulnerable to destructive high-intensity wildfires. Decades of fire suppression only postponed the inevitable.

In the summer of 2000, 90 years after fires encircled Ed Pulaski and his men, the West was again burning. Lightning storms ignited a series of blazes in Montana's Bitterroot Valley. In a single five-day period more than sixty fires began. Crews responded quickly but could bring only a couple dozen of them under control. The others continued to burn, driven more by extremely dry conditions and high temperatures than by the wind. Veteran firefighters faced extremes they had never seen before. Walls of flame 150-feet high were spreading across thousands of acres an hour.

Early that May the fire season had burst into the headlines when an attempt to reduce the wildfire hazard at Bandelier National Monument created the very disaster it aimed to prevent. Crews purposely set a prescribed fire to reduce dangerous accumulations of vegetation, and then the winds picked up. They set a backfire to keep it from spreading

across the national forest toward the town of Los Alamos. But strong winds drove the flames into thick stands where the fire climbed into the forest canopy. A firestorm raced through the treetops, throwing firebrands into the community. Eventually 237 homes disappeared in flames spread by surface fires and falling embers. The tremendous impact on human lives highlighted a major change that had occurred in the West since 1910. Millions of people now lived in regions prone to burn.

The 2000 fire season reached peak intensity on August 29 when more than 28,000 firefighters faced scores of large fires burning out of control in 16 states. They were supported from the air by more than 200 helicopters and 42 air tankers, dropping water and fire retardant to slow the spreading flames. Thousands of soldiers joined crews from 38 states and 8 Indian tribes. With resources stretched to the limit, extra crews and management teams flew in from Canada, Mexico, New Zealand, and Australia to join the effort. Firefighting had gone international. By the time rain and cooler temperatures arrived in September, the fires in the Bitterroot area had consumed 356,000 acres and burned 70 homes.

When the change of seasons finally brought an end to the fires, forest managers began totaling the staggering losses. Nationwide, more than 7 million acres burned, 577 structures were destroyed, and 20 firefighters lost their lives. It was one of the worst fire seasons on record, and even more troubling, a sign of what the future might hold.

By the beginning of the twenty-first century wildland fires were occurring more frequently, killing normally fire-resistant old-growth stands and destroying the soil nutrients essential to forest recovery. The growing threat of extreme fires spurred governments and private organizations to take a closer look at the nature of fire, how to fight it, and how to live with it.

fuel, terrain, and weather

FIRE CAN SURPRISE THE MOST EXPERIENCED CREWS. At times it appears to have a mind of its own, sputtering along the ground for hours until suddenly flaring into the crowns of trees with a deafening roar. It can take unexpected turns, leaving some places unscathed while obliterating others. But those whose lives depend on understanding the mercurial nature of fire focus on certain tangibles: fuel, terrain, and weather. Any organic matter capable of burning is considered a fuel. The more fuel available, the more intense the fire. Fire races through light fuels such as grasses and brush, spreading fast and quickly exhausting its source. Heavy fuels, including snags and windfall, are slow to ignite but can sustain a fire for longer periods. A burn often consumes only a small amount of the vegetation, usually the finer fuels, leaving many trees only scorched or charred.

If the moisture content of the fuel drops low, the likelihood of a fire grows. When moisture content is high, a fire is unlikely to start and will burn poorly if it does. Rangers regularly check this during the fire season to predict the potential for combustion, and track it during an actual fire to help determine how hot and fast the fire will burn. Instruments and computer modeling help forest managers predict the chance of a fire igniting, but experienced firefighters develop a feel for it. "If you've spent enough time in the woods," said forest ranger Jeff Riepe, "you can smell it."

RIGHT: Flames leap to treetops, causing a deep black plume to rise from an Idaho forest in the summer of 2000.

ABOVE: Rachel Longknife of the
Globe Hotshots of Arizona cools off
an advancing fire with a backpack
pump or "bladder bag."

ABOVE: Tony Sanchez, squad boss on the
Payson Hotshots, uses a fusee (flare) to
burn fuel along a fireline. This arson-
caused blaze was on the Salt River Indian
Reservation in Arizona.

LEFT: Wildfire smoke obscures the sun over
Sandy, Utah, a suburb of Salt Lake City.

Terrain features play an important role in shaping wildfire. Elevation, the angle of a slope, the direction it faces, and the terrain complexity all influence fire. The steeper the slope, the faster fire will move up it. South-facing slopes, absorbing more direct sunlight, dry faster and heat quicker. Canyon and mountain environments, with high vertical relief and intricate drainage systems make fire behavior harder to read.

Of all the elements of a fire, weather is the most difficult to predict. Winds can intensify and shift direction, causing a sudden blowup and endangering the lives of those caught unprepared. Firefighters keep a constant watch on the weather for any sign of deteriorating conditions. Their safety depends on knowing if the weather will trigger a sudden change in the fire's direction, rate of spread, or intensity. They measure and record the relative humidity, knowing that when it is low, the danger of fire is high. An increase in humidity or precipitation can slow a fire or completely extinguish it. Normal seasonal drying and prolonged drought increase the risk of wildfire, and the combination of both can create extreme fire conditions.

During extreme fire season

RIGHT: A groundfire leaves trees charred but alive and prevents buildup of fuel.

Wildland firefighters pay attention to air temperature since a fire becomes more active as the day heats up. And to determine the degree of atmospheric stability, they notice cloud types, the motion and color of smoke, and any change in winds. A sudden calm often precedes a downdraft and explosive conditions. In unstable weather, a low-intensity fire can transform into a dangerous blowup with little warning.

Wind is the most critical factor influencing fire. It affects how fast the fire spreads and the direction it moves. It can increase evaporation rates, drying the air and the fuels, increase the rate of combustion by supplying more oxygen, and preheat timber in advance of a fire front. Winds also carry firebrands and embers long distances, sometimes crossing firelines and starting spot fires. The dynamics of a large fire will create its own weather. A rising column of heat sucks air into it from all directions and creates winds that can reach hurricane force, ripping full-grown trees from the ground.

Positioned throughout remote areas of the United States, automated weather stations collect information, including wind speed and direction, precipitation and air temperature, and relative humidity and fuel moisture. Twice a day the National Weather Service issues fire weather forecasts. When conditions indicate a growing threat of fire, they post a "Fire Weather Watch." A day or two later they will issue a "Red Flag Warning" if the fire weather reaches a critical stage. Lightning-generating storms, especially those producing little or no rain, cause the greatest number of wildland fires. Each year lightning ignites more than 10,000 fires in the United States.

In 1988 TV viewers watched images of the country's oldest national park going up in smoke. More than 240 fires roared through the Yellowstone region of Wyoming, destroying vast stands of lodgepole pine. Lightning ignited most of them; others were human-caused. Smaller fires burned together, creating several major conflagrations with walls of flame reaching 200-feet high and smoke columns rising 30,000 feet into the sky. For weeks, thousands of firefighters battled to control the blazes until the change of seasons finally brought an end to the long, hot summer.

Rick Gale, former chief of fire and aviation for the National Park Service, served as Area Commander during those fires and knows there are times when they can only do what the fire lets them do. "During extreme or severe fire seasons," he said, "all the firefighting resources in the free world aren't going to make one whit of difference."

Two major lightning fires hit Mesa Verde National Park in Colorado in 2000. Fire crews, guided by archeologists, protected threatened prehistoric sites such as Long House, the second largest cliff dwelling in the park. Ultimately, only a few archeological sites suffered smoke or heat damage, but the fires charred nearly 40 percent of the park, burning stands of Utah juniper, piñon pine, and Gambel oak. The oak thickets will recover within a few years, but the piñon-juniper woodlands may take centuries to regenerate. The second fire burned large tracts in the adjacent Ute Mountain Ute tribal park, destroying the historic cabin of Chief Jack House, the last traditional leader of his band.

ll the firefighting resources in the free world
aren't going to make one whit of difference.

Fires can also be ignited by a careless smoker, a campfire left unattended, and sometimes—by arsonists. The Jasper Fire ripped through the Black Hills of South Dakota the same summer Mesa Verde burned. It swept over most of the surface at Jewel Cave National Monument. An established program of thinning and prescribed fire deprived the wildfire of fuels, slowing its spread and giving crews time to save the historic buildings. Investigators determined that someone had set the fire intentionally.

a fire-shaped land

FIRE IS A MAJOR FORCE IN SHAPING THE WESTERN LANDSCAPE. Most North American ecosystems benefit from fire and some depend on it, since fire recycles nutrients essential for the growth of many plant species. Once a tree is reduced to ash, rain dissolves the nutrients and returns them to the soil. By vigorously suppressing fire, land managers have altered the natural balance of plant and animal communities. "In many cases," said Steve Holder, Interagency Safety Program Manager, "fire sustains and shapes major ecosystems."

A mosaic of different vegetation types appear after a fire. The flame-swept landscape slowly changes as one species of plant replaces another. Pioneering aspen and lodgepole soon take root, only to be replaced by a more diverse forest. A variety of plant communities reduces the chances of disease destroying an entire forest. "The forest ecosystem survives because of this diversity," said Dennis Knight, a forest ecologist at the University of Wyoming, "and maintaining that diversity depends on periodic disturbance."

More than 100 years of excluding fire has altered the natural configuration of the forest, making it more susceptible to destructive wildfire. A recent study found 39 million acres of the nation's forests were at a high

RIGHT: A new fire tints the evening sky.

risk of fire. Conditions deteriorate as deadwood accumulates and stunted, crowded trees compete for sunlight and water. These fuels feed more severe fires that burn hotter, spread faster, and cost more to suppress. Intense fires can sterilize the soil, reducing the chances for the original biological diversity to recover.

The unique western landscape reflects thousands of years of interaction among fire, forests, and human activity. Humans are part of the natural history of these forests. In many areas, Paleo-American hunters were on the scene from the beginning when trees began to colonize vast regions at the end of the last ice age. European settlement during the nineteenth century introduced a new set of conditions—the grazing of livestock, logging for construction, woodcutting for fuel, and fire suppression. These activities disrupted the established cycle of fire, a pattern that varies with terrain and elevation.

xcluding fire has altered the nature of the forest, making it more susceptible to destructive wildfire.

"Other than the change of seasons," said fire historian Tom Swetnam, director of the University of Arizona's Laboratory of Tree-Ring Research, "fires used to be the most frequent and important natural event in ponderosa pine forests."

Forests throughout the West have adapted to fire in different ways. Certain types of fire can benefit a forest, while other types can destroy it. The dryland ponderosa forests, found in the Southwest and the lower elevations of the Rockies, thrive with frequent, low-intensity surface fires. Ponderosa and Douglas fir have developed thick, heat-resistant bark to survive ground fires. In the past, these surface fires flared up every two to seven years in some locations and ten to twenty years in others. Ground-hugging flames less than three-feet high swept through stands of thick-bark pines doing little damage to them while clearing out the understory vegetation. Within several years the forest returned to a prefire stage.

LEFT: A giant smoke plume dwarfs a helicopter dragging a bucket.

21

Eliminating surface fires in these forests disrupted the equilibrium, allowing thickets of dog-hair pine to grow unchecked and surface fuels to accumulate. This buildup of fuels set the stage for more destructive fires. A fire now climbs the dead and dying timber into the canopy, running from treetop to treetop in a devastating crown fire. Once in the canopy, the fire burns intensely and flames shift direction unpredictably. As it gains energy, fire crews find themselves facing a firestorm. It will destroy pinecones before they reach maturity and release their seeds, leaving the forest without a means to regenerate. A ponderosa forest can take centuries to recover from a crown fire.

Northern, higher-elevation forests of lodgepole pine and Engelmann spruce have adapted to stand-replacing fires. Lodgepole pines not only tolerate crown fires but also need them. Their cones hang from the top branches for years until a crown fire ignites. Only intense heat will cause them to open and drop their seeds. Stand-replacing fires are a natural part of the ecosystem, so a build-up of fuels is not as critical a factor as it is in lower-elevation forests.

Knight added, "I believe quite strongly that the fires of 1988 in the Yellowstone area were due primarily to climatic conditions, not fuel accumulations caused by fire suppression. Fire suppression and fuel accumulation probably had a significant impact at lower elevations where frequent, low-intensity fires were the rule, but such forests cover a small area in Yellowstone. At higher elevation where less-frequent high-intensity fires prevailed in the past, climate seems to have been a more important factor."

Even before aggressive firefighting became policy, the frequency of fire in certain western forests abruptly decreased. Herds of sheep and cattle began grazing on the grassy understory, removing the light fuels capable of carrying the fire close to the ground. Brush and saplings invaded the openings, having much the same effect as the policy of fire suppression with parks filling in and meadows becoming overgrown.

Grazing and fire exclusion changed the face of the forests. In 1881 the ponderosa forests of northern Arizona had little undergrowth with widely spaced trees growing at a density of 56 trees per acre. A century later hundreds, and sometimes thousands, of trees crowded each acre.

ABOVE: A firefighter rushes to the next burn in the Bitterroot fires of 2000.

BELOW: Union Hotshots down, cut up, and quench flaming trees to prevent the fire from spreading through crowns.

RIGHT: Pattern left by a wildfire on Bunsen Peak in Yellowstone National Park.

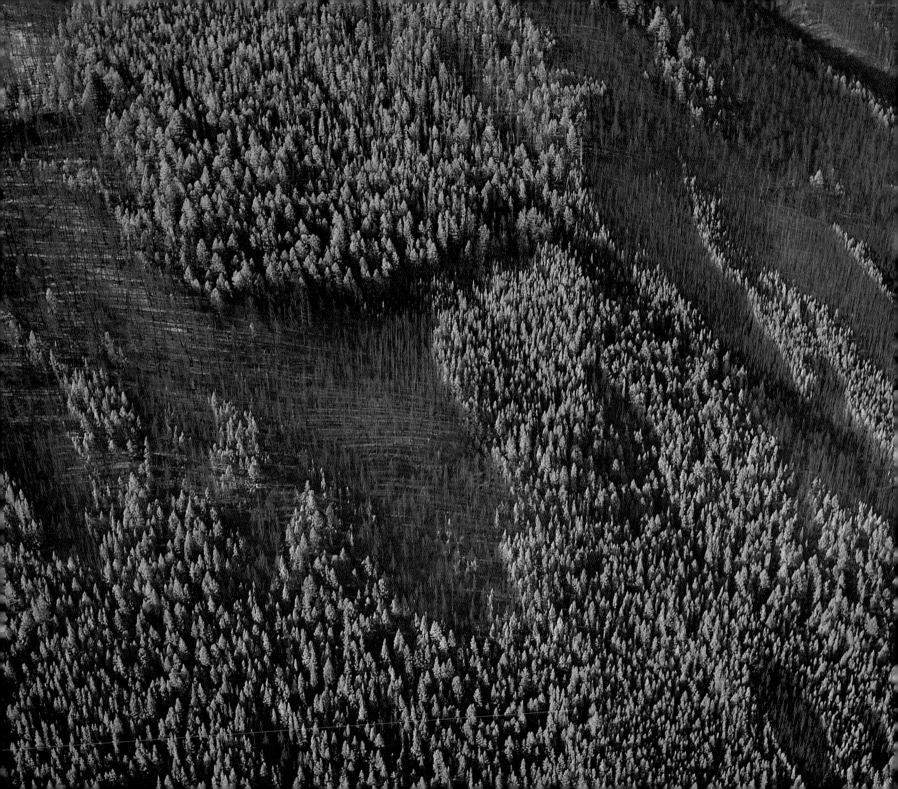

indian flat

DOWNWIND OF THE FIRE, THE AIR SMELLED BURNED. Flames stretched in a ragged front across Indian Flat as a massive column of smoke rose in the air and leaned above a crew of firefighters. They spread out in a skirmish line among the junipers and piñons, clearing a fuel break. A long drought had parched the land, leaving the trees drier than a stack of kiln-dried wood. A ponderosa in the distance burst into flames sending up a black plume. And then another, and another.

On June 20, 1996, lightning struck in the Hochderffer Hills north of Flagstaff, igniting the blaze. Before ending, it would burn 16,350 acres, making it the largest fire ever recorded in the Coconino National Forest.

RIGHT: A brush truck relocates quickly after the wind shifts during a firing operation. The engine was part of a five-truck strike team relocated to Arizona from Idaho during the busy desert fire season. Later in the year, resources from the Southwest will make the trip north to help fight fires there. Sharing resources like this is one of the strengths of the interagency dispatch system.

RIGHT: Firefighters evacuate a forward camp during the Storm Creek fire in Yellowstone National Park.

Thick smoke billowed skyward half a mile from where John Farella worked to fire-proof his home, cleaning away the brush and lumber as a precaution. He was out of the direct path of the fire, and with the winds blowing steadily to the east he thought it would miss him. The relative humidity stood at four percent; gusts were hitting 45 mph. And then the winds shifted. An arm of the fire veered north and began a run straight toward the homestead. As embers dropped from the dark sky, John climbed in his truck and left the house in the hands of the firefighters. "You couldn't help but be afraid," he said, "the way it was blowing."

A forest service pumper with a crew of three took shelter behind the stone house; a fire engine was stationed in front with four firemen. They had foamed the structure and now prepared to defend it. Two forest service firefighters, Don Muise and Chris Bradley, took positions on the metal roof. Wearing yellow flame-resistant shirts, they worked a hose and waited for the fire to hit.

Across the field, safety officer Lee Kimball watched as flames swept over the ridge toward the house. The fire was moving fast, generating tremendous heat. At a certain intensity, what a fire consumes no longer matters, and under the right conditions, it develops its own momentum. Instead of spreading slowly outward like a stain, it finds a direction and runs. "A fire creates its own energy," Lee said, "its own winds. Wind and topography determined the path of this fire, not fuels."

The fire closed the distance, burning with such intensity entire trees were being incinerated to nothing but ash. The two men exposed on the roof watched the fire rush toward them with a roar. "We can do this," Chris told his partner, "we can do this."

On the ground next to the house the rest of the crew crouched low and braced for the shock. They thought the men exposed on the roof must have taken shelter, but they stayed put, determined to ride it out. The head of the fire crashed down the hill and hit a stand of ponderosa behind the house, exploding into a 100-foot wall of flame. Suddenly it was arching above the firefighters and leaping over the roof. The flames raced across the open field beyond, leaving the house standing intact on an island of unburned ground. Everything around it was charred and blackened. "Those guys had a wall of fire roll over them," John said. "It was a firestorm."

When fire crews do their job, it doesn't make headlines. That evening the news described a fire still raging out of control, gave the number of acres burned, and then filled in the report with what the army of 1000 firefighters had eaten for breakfast that morning: 3000 eggs, 250 pounds of bacon, and 250 gallons of coffee. Acts of courage get lost in the numbers.

27

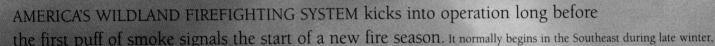

chasing smoke

AMERICA'S WILDLAND FIREFIGHTING SYSTEM kicks into operation long before
the first puff of smoke signals the start of a new fire season. It normally begins in the Southeast during late winter, then shifts to the Southwest in spring, and finally moves north through the Rockies to California and the Pacific Northwest. Crews have been hired and trained, supplies readied, equipment moved to forward staging areas. Automated weather stations have been gathering critical data and transmitting it via satellite to the national fire center. The weather service has been making daily forecasts, and the fire intelligence office

A lightning-caused wildfire burns grass and
sagebrush along a highway in Nevada.

Small fires are hit early durin

Fires often begin as a single burning snag with flames at its base. Idaho smokejumper Bobby Montoya once parachuted into a fire along the remote Middle Fork of the Salmon River. Lightning had struck the top of a large tree, and the standard procedure was to fell it with a chainsaw, buck it up, and then cool it down with dirt and water. Told to cut it down, Montoya looked at that magnificent old tree and decided to disregard his instructions. He climbed the 140-foot tall pine and spent hours hauling up water to put it out. "That tree," he says with pride, "is still there."

reme conditions to prevent them from growing into runaway fires requiring a tremendous effort to fight and increased risk.

has been pooling reports from around the country. Air tankers and helicopters are standing by at airfields. Patrols have been driving the backroads, watching for smoke and warning visitors about the danger. Smokey the Bear posters have begun appearing on signboards. And the fire crews, on call 24 hours a day, wait.

When the first smoke appears, a fire lookout or aircraft flying overhead soon detects it. They call in a report to the dispatcher by radio. The fire dispatch office receives it, rapidly assesses the threat, and sends out the initial attack force. Small fires are hit early during extreme conditions to prevent them from growing into runaway fires requiring a tremendous effort to fight and increased risk.

Airborne firefighters, either smokejumpers or a helitack team, normally handle fires breaking out in remote areas. Helitack crews land near a burn or rappel from a hovering chopper, reaching the ground in 20 seconds. Smokejumpers, working in small units, parachute into inaccessible locations and quickly attack the blaze. If the fire is within 100 miles of their base, they will be on the scene within half an hour. They jump two at a time, followed by a cargo drop of tools and food to last 48 hours. When the job is done, they pack out to the nearest trailhead or pickup point, carrying up to 110 pounds of gear on their backs.

Once the fire is spreading along the ground, a hotshot crew handles the initial response. Highly trained, these 20-person teams take on many of the toughest duties, using a variety of specialized tools to build firelines, set backfires, and mop-up. They drive as close to it as possible and then bushwhack to ground zero or get a lift in by helicopter. If the fire is near a road and in light fuels, a wildland engine crew might handle the initial attack. These trucks, designed for backroad use, carry up to 800 gallons of water. The actual response depends on the nature of the fire, the access to it, and the availability of crews to fight it.

No matter how these smokechasers get to work, their job is to keep the fire from spreading. Hand crews do this by building a fireline, a cleared fuel break around the perimeter of the fire to remove any burnable material. They first cut down the trees and brush, followed by scraping down to mineral soil. If the situation warrants, they will burn out the vegetation between the line and the fire. A sawyer roughs out the line with a chainsaw, followed by firefighters turning the dirt with a shovel or Pulaski, a hybrid tool with an ax on one side of the head and a hoe on the other, named to honor the hero of the 1910 fire.

on the fireline

ON INTENSE FIRES, CREWS BEGIN CUTTING
FIRELINE in the rear and work along the sides
to pinch off the head. They also use a drip
torch for igniting backfires or burnouts. Once
the line is in, the next job is to keep the fire
from jumping it. When heavy equipment can reach the site, tractor-plows and bulldozers
clear a "Cat line." Other crews foam historic cabins or wrap them in
heat-resistant foil as protection from firebrands.

Many firefighters are pulled from their normal duties as park and forest
rangers and organized into crews when major fires require all hands.
Breathing smoke and facing intense heat, these men and women spend
long hours doing difficult and sometimes dangerous work. They may
have to bivouac overnight without sleeping bags or hot meals.
"Sometimes," said firefighter Michael Payne, "you have to coyote it."

During the initial stages of a fire, the incident commander may call in
an aerial attack. Air tankers lumber in, delivering up to 3000 gallons
of chemical fire retardant or water to slow the fire. This gives ground
crews time to build firelines. Lead planes guide the air tankers on their
runs and perform aerial reconnaissance. And when a fire directly
threatens life and property a C-130 aircraft can release a pressurized
3,000-gallon tank in a single burst. Helicopters also deliver loads of
water or retardant from a suspended bucket, useful for hitting spot
fires and plugging holes in a fireline. The largest chopper, a Sikorsky
Sky Crane, can douse a hot spot with 2000 gallons.

RIGHT: Suppressing a grass and
brush fire near the Boise River.

ABOVE: Two attempts to stop wildfire: digging a dirt fireline and burning fuel in the fire's path.

BELOW: A firefighter drenched in foam.

LEFT: Air tanker 99 drops retardant to slow a fire's advance.

As the fire-fighting effort grows, additional support is mobilized. Helicopters transport people and carry supplies in slings belly-hooked to the chopper. They also drop aerial incendiaries if the incident commander needs to fight fire with fire. Aircraft equipped with infrared scanners locate hot spots as small as six inches in diameter from an altitude of 8000 feet. They fly at night when a wider temperature difference between the ground and fire increases the chance of locating a heat source. Precise mapping of the fire's extent allows commanders to develop the right strategy to suppress it.

On the rare occasions when an initial attack is unsuccessful, the incident moves into an extended attack phase. Land-management agencies dispatch additional resources as needed, calling on other agencies for help. If the agencies within a region run short of firefighters and equipment, a request is forwarded to the nearest of eleven regional coordination centers. These centers find what is needed and fly it in. When they are unable to locate the necessary resources within the region, the request goes to the national coordination center, an arm of the National Interagency Fire Center in Boise, Idaho. "Fire," said Steve Holder, "doesn't recognize boundaries between agencies."

Involvement on the national level depends on the size and the complexity of the fire. When the need exists, they can dispatch an army of fire crews, management teams, aircraft, and supplies to any region of the country. They can also request assistance from the military. In the summer of 2000, resources were stretched to the limit. Soldiers and Marines assisted on the fireline, and for the first time the national fire center asked for overseas aid.

Fighting a major wildland fire requires the movement of thousands of people, the organization of immense quantities of equipment and supplies, and the coordination of tactical air support. A command center sets up at the scene of the fire to manage the complex logistics and the 24-hour suppression efforts. The work is grueling, both on the fireline and at the command level. "Keeping people safe," said Paul Broyles, National Park Service fire operations manager, "is our first priority. Working up to sixteen hours a day for two weeks, the stress and fatigue builds up."

Literally overnight, a fire camp appears. Showers are installed, a dining hall opens, a medical tent is pitched. Firefighters have mail delivered to them and access to banking services. The camp fills with food caterers and water-hauling contractors, heavy equipment operators, bus drivers to shuttle the fire crews, law enforcement, and even a public relations staff to handle the media. "It's amazing," said Michael Payne, who has run these base camps. "Within 24 hours we can put up a city in the middle of a meadow."

Out on the fireline, crews work either day or night shifts. It is exhausting work and potentially dangerous. Most wildfire deaths happen when flames overrun fire crews or cut off their escape. Surprisingly, a majority of entrapments happen on the smaller fires or an isolated sector of a big fire. A blowup occurs when wind, dryness, and terrain combine to create a deadly situation. The fire suddenly flares up as winds shift or increase speed. Once in the canopy, it spreads rapidly, generating intense heat and walls of moving flame. A crown fire cannot be contained. "At a certain level of fire intensity," said Sue Vap, national fire management officer, "we're not in charge."

When a fire is about to overrun them, firefighters deploy their fire shelters. They have been trained to shake out and climb into these foil tents in less than 30 seconds. In 1985 a blowup at the Butte Fire in Idaho trapped 73 firefighters who were forced to deploy their shelters. Investigators later determined that 60 of them would have died if they had not used them.

A sense of shared danger forms a bond among smokejumpers. Tacked to the wall of the smokejumper loft in Missoula is a poster for Young Men and Fire, Norman McLean's tragic story of the Mann Gulch Fire where thirteen smokejumpers lost their lives in 1949. Down the hallway, a wall plaque honors Don Mackey, a Missoula jumper who died on Storm King Mountain in 1994. That fire trapped experienced firefighters from smokejumper, hotshot, and helitack teams. The incident took a fatal turn in an amazingly short span of time. The blowup happened at 4:00 in the afternoon. They recognized the danger at 4:10, and by 4:13 the flames had overrun and killed fourteen firefighters.

Showers are install

"Those guys about to die," said Bobby Montoya, who fought wildfires for twenty years, "I know what they're thinking. They can't breath it's so hot. The fire is sucking the air around you—your clothes and hair start vibrating with the air rushing by you. There's no oxygen. The fire is sucking it out of you. At the last moment they're not thinking about fire; they're thinking about air."

But a blowup isn't the only danger a firefighter faces. "I've had closer calls at night from falling snags and rocks," said Montana smokejumper Walt Smith. "You're usually in steep terrain where the rocks are black and the ground is black. You can't see them coming. All you hear is 'thump, a-thump, thump,' and you don't know which way to run."

Large mammals rarely get trapped by fire. They hoof it to safety outside the perimeter or to refuge areas within the burn. Burrowing animals will duck and cover, taking shelter below ground. But rabbits and other small, surface-dwelling animals, such as the raccoon and porcupine, often suffer high losses. Birds can escape an approaching fire on the wing, although nesting areas may be destroyed. One species of bird, the Kirtland's warbler, colonizes only recently burned stands. A fire shuffles the deck, resulting in new habitats and a shift in species.

Mop-up operations begin when a fire is fully contained. Crews crisscross the burned ground, separating hot material from cold. With backpack pumps made from collapsible neoprene, they cool the embers and hit each smoke until the fire is extinguished. To be certain, they "cold trail" the burn, running their bare fingers through the ash to detect hot spots.

Literally overnight, a camp appears.
dining hall opens, a medical tent is pitched.

bear jaw

BEAR JAW CANYON ANGLED UP THE MOUNTAIN IN A STEEP CREASE, choked with heat-reddened rocks and burned pines. A Navajo medicine man walked up it with his assistants. They had come to the San Francisco Peaks to perform a healing ceremony after fire had swept up their sacred mountain of the west. Finding the point where the fire began, the traditional singer tied on a black ceremonial headband and draped a multi-strand turquoise necklace over his bola tie. He spread a blanket on the ground and opened an attaché case filled with ceremonial paraphernalia. Next, he arranged a collection of neatly tied buckskin pouches, representing the four sacred mountains, as his apprentice prepared a ritual smoking mixture of native tobacco, other herbs, and shavings from the horn of a bighorn sheep. When all was ready, the ceremony opened. A rhythmic singing began softly, the lips of the medicine man barely moving. His voice had a distant quality to it, as if the song was coming from somewhere faraway.

LEFT: Stumps and downed trees continue to pose a fire threat.

"We do this to put out the black fire,

Clouds and smoke blend over the
Bitterroot National Forest, Montana

"The first song," he explained, "goes to the home in the sky, the black clouds, the male rains. Then the seas, the home of water, the female rains . . ." He passed a pouch of corn pollen to those sitting in a circle. Following his example, each took a pinch and touched the tongue, then above the forehead, ending with a sprinkle toward the ground and a sweeping motion skyward.

When the singer was ready for the cooling-down phase, he asked Linda Farnsworth, forest service archeologist, to assist him. "We'll make a Navajo out of you," he joked and handed her a pair of feathered prayer-sticks. "Sprinkle it like this." He took her hands and showed her the proper way to shake them. "There you go. You represent the female rain."

the medicine man said,
"the blue fire, the yellow fire, the white fire."

His apprentice, representing the male rain, walked around the circle and sprinkled each person from the sky down as Linda followed, sprinkling from the ground up. They made four rounds and every pass felt as cooling as a rain shower on a summer day. "We do this to put out the black fire," the medicine man said, "the blue fire, the yellow fire, the white fire."

Ashes from the ritual smoke were passed around and rubbed on the hands for protection. And then the Navajo singer brought the ceremony to a close with a prayer "for everlasting beauty."

renewal

WHEN THE FIRES HAD COOLED on Mesa Verde in the late summer of 2000, two American Indian groups arrived to perform ceremonies. Already on the scene was an emergency rehabilitation team, assessing the damage done by the fire and the act of suppressing it.

Within days they reported on immediate measures to protect against sheet erosion, reduce threats to rare species, and stabilize archeological sites. The fires exposed more than 1000 sites, ranging from scatters of pottery to mounds of rubble where ancient walls had collapsed. Many of them, previously hidden by the vegetation, had never been recorded. Crews quickly worked to cover those most threatened with a stabilizing mesh and to build drainage. Later they reseeded the area with native grasses to control erosion and prevent the invasion of non-native plants. An attempt is made to restore the landscape to its prefire appearance, but with intensely burned stands of piñon-juniper it may take up to 300 years for the woodland to return.

At first the damage to Yellowstone after the 1988 fires appeared catastrophic. More than a third of the park was reduced to ash and charred trees, but it rebounded quickly. After the fires, a researcher found a million lodgepole seeds released in a single acre, and a dozen years later seedlings covered much of the burned-over landscape.

Tom Swetnam, who studies the history of fire recorded in tree rings, testified before Congress on the devastating fires of 2000. "For those who live in the West the situation is clear," he said. "Wildfires will continue to threaten communities and destroy more homes. The only question is when. Fires will inevitably occur when we have ignitions in hot, dry, windy conditions. If there is fuel to burn, it will burn intensely. It is one of the great paradoxes of fire suppression that the more effective we are at fire suppression, the more fuels accumulate and the more intense the next fire will be. We must learn to live with this reality."

Land managers in the West began a shift from fire control to active fire management in the 1970s. They reintroduced fire into ecosystems to reduce the buildup of dead trees and lessen insect infestations, thinned stands, and sometimes treated fuels by biological and chemical means. Park rangers at Theodore Roosevelt National Park have used insects to control knapweed, an exotic species. Of these measures, the greatest change in thinking was the use of prescribed fire—a low-intensity surface fire intentionally ignited to reduce the chance of a catastrophic wildfire.

Surface fires kill few large trees and stimulate the growth of grasses and shrubs that protect the watershed. They create open stands and improve habitats by recycling nutrients and improving forage for wildlife. A lightning strike becomes a prescribed fire when allowed to continue burning under strict conditions. Whether a fire is purposely ignited or a natural fire allowed to burn, both are kept within narrow boundaries. Land managers start them only under certain weather conditions when the behavior of the fire can be controlled. They need the right amount of wind to disperse smoke but not enough to let the fire spread outside the limits of the burn. They also consider the impact on such things as water and air quality, visibility, cultural resources, and wetlands. Written plans for a managed fire must be approved in advance, and since fire is inherently dangerous they have contingency plans if it burns out of control.

Some forests and brushlands don't lend themselves to prescribed fire. In the northern lodgepole forests, a ground fire is likely to climb the staircase of branches and become a crown fire. When it's safe to burn in the chaparral-covered foothills of southern California, it is generally too wet. And a prescribed fire can easily escape by the time the brush

Wildfire pushed into the wooded community, threatening the nation's nuclear laboratory, destroying houses, and disrupting the lives of thousands who were forced to evacuate their homes. Some observers saw the tragedy as an indictment of the policy of using prescribed fire to thin the forests. Others viewed it as warning of things to come if land managers did not undertake more prescribed burns to restore the natural cycle of fire.

Wally Covington, a forest ecologist at Northern Arizona University, believes high-intensity fires can do more ecological damage than clear-cutting a forest. He is in the forefront of those who advocate the restoration of forests using a combination of thinning smaller trees and then setting prescribed fires. Hand crews with chainsaws normally do the thinning. They remove the cut wood with a trailer pulled by an all-terrain vehicle, reducing soil compaction and disturbance. But Covington's approach remains controversial. Some environmental groups believe this type of intervention requires too much logging and will actually increase the chance of fire, while others hold the view that the cutting of any trees cannot be justified. Lawsuits have been filed; accusations fly back and forth. Meanwhile, the fires continue to burn.

Bobby Montoya began smokejumping in 1962 and has seen firsthand how the forests have changed. "When I started this," he said, "it was an adventure. We were all Amelia Earharts—we wanted the adventure. In that airplane there were fifteen of us and we were undefeated. Nothing was going to stop us. Fire was the enemy. All those fires we fought just came back on us. We did too good a job on it, and I've lived to see it happen." He paused, thinking back on twenty seasons of fighting fires. "This country has always burned—and it needs to burn."

Surface fires kill few large trees and stimulate the growth of
They improve habitat

reaches a burnable stage. Most land managers agree on the necessity of reducing extreme wildfire conditions, but disagree over the right mix of natural fires, thinning, and prescribed fires to accomplish this. The complexity and inherent danger only grows when a community borders the forest. Along the wildland-urban interface, the stakes are high, as graphically illustrated by the Cerro Grande Fire at Los Alamos.

RIGHT: The Swan Flats area of Yellowstone National Park shows regeneration years after the fires of 1988.

rasses and shrubs that protect the watershed.
y recycling nutrients and improving forage for wildlife.

All those fires we fought just came back on us